SHOE SIZING and FITTING

An analysis
of practices
and trends •

Miscellaneous Publication No. 469
U. S. DEPARTMENT OF AGRICULTURE

A History of Shoemaking

Shoemaking, at its simplest, is the process of making footwear. Whilst the art has now been largely superseded by mass-volume industrial production, for most of history, making shoes was an individual, artisanal affair. 'Shoemakers' or 'cordwainers' (cobblers being those who repair shoes) produce a range of footwear items, including shoes, boots, sandals, clogs and moccasins – from a vast array of materials.

When people started wearing shoes, there were only three main types: open sandals, covered sandals and clog-like footwear. The most basic foot protection, used since ancient times in the Mediterranean area, was the sandal, which consisted of a protective sole, attached to the foot with leather thongs. Similar footwear worn in the Far East was made from plaited grass or palm fronds. In climates that required a full foot covering, a single piece of untanned hide was laced with a thong, providing full protection for the foot, thus forming a complete covering. These were the main two types of footwear, produced all over the globe. The production of wooden shoes was mainly limited to medieval Europe however – made from a single piece of wood, roughly shaped to fit the foot.

A variant of this early European shoe was the clog, which were wooden soles to which a leather upper was attached. The sole and heel were generally made from one piece of maple or ash two inches thick, and a little longer and broader than the desired size of shoe. The outer side of the sole and heel was fashioned with a long chisel-edged implement, called the clogger's knife or stock; while a second implement, called the groover, made a groove around the side of the sole. With the use of a 'hollower', the inner sole's contours were adapted to the shape of the foot. In even colder climates, such designs were adapted with furs wrapped around the feet, and then sandals wrapped over them. The Romans used such footwear to great effect whilst fighting in Northern Europe, and the native Indians developed similar variants with their ubiquitous moccasin.

By the 1600s, leather shoes came in two main types. 'Turn shoes' consisted of one thin flexible sole, which was sewed to the upper while outside in and turned over when completed. This type was used for making slippers and similar shoes. The second type united the upper with an insole, which was subsequently attached to an out-sole with a raised heel. This was the main variety, and was used for most footwear, including standard shoes and riding boots.

Shoemaking became more commercialized in the mid-eighteenth century, as it expanded as a cottage industry. Large warehouses began to stock footwear made by many small manufacturers from the area. Until the nineteenth century, shoemaking was largely a traditional handicraft, but by the century's end, the process had been almost completely mechanized, with production occurring in large factories. Despite the obvious economic gains of mass-production, the factory system produced shoes without the individual differentiation that the traditional shoemaker was able to provide.

The first steps towards mechanisation were taken during the Napoleonic Wars by the English engineer, Marc Brunel. He developed machinery for the mass-production of boots for the soldiers of the British Army. In 1812 he devised a scheme for making nailed-boot-making machinery that automatically fastened soles to uppers by means of metallic pins or nails. With the support of the Duke of York, the shoes were manufactured, and, due to their strength, cheapness, and durability, were introduced for the use of the army. In the same year, the use of screws and staples was patented by Richard Woodman. However, when the war ended in 1815, manual labour became much cheaper again, and the demand for military equipment subsided. As a consequence, Brunel's system was no longer profitable and it soon ceased business.

Similar exigencies at the time of the Crimean War stimulated a renewed interest in methods of mechanization and mass-production, which proved longer lasting. A shoemaker in Leicester, Tomas Crick, patented the design for a riveting machine in 1853. He also introduced the use of steam-powered rolling-machines for hardening leather and cutting-machines, in the mid-1850s. Another important factor in shoemaking's mechanization, was the introduction of the sewing machine in 1846 – a development which revolutionised so many aspects of clothes, footwear and domestic production.

By the late 1850s, the industry was beginning to shift towards the modern factory, mainly in the US and areas of England. A shoe stitching machine was invented by the American Lyman Blake in 1856 and perfected by 1864.

Entering in to partnership with Gordon McKay, his device became known as the McKay stitching machine and was quickly adopted by manufacturers throughout New England. As bottlenecks opened up in the production line due to these innovations, more and more of the manufacturing stages, such as pegging and finishing, became automated. By the 1890s, the process of mechanisation was largely complete.

Traditional shoemakers still exist today, especially in poorer parts of the world, and do continue to create custom shoes. In more economically developed countries however, it is a dying craft. Despite this, the shoemaking profession makes a number of appearances in popular culture, such as in stories about shoemaker's elves (written by the Brothers Grimm in 1806), and the old proverb that 'the shoemaker's children go barefoot.' Chefs and cooks sometimes use the term 'shoemaker' as an insult to others who have prepared sub-standard food, possibly by overcooking, implying that the chef in question has made his or her food as tough as shoe leather or hard leather shoe soles. Similarly, reflecting the trade's humble beginnings, to 'cobble' can mean not only to make or mend shoes, but 'to put together clumsily; or, to bungle.'

As is evident from this short introduction, 'shoemaking' has a long and varied history, starting from a simple means of providing basic respite from the elements, to a fully mechanised and modern, global trade. It is able to provide a fascinating insight not only into fashion, but society, culture and climate more generally. We hope the reader enjoys this book.

Contents

	Page
Toward better fitting footwear	1
Background of present practices	3
How last manufacturers arrive at size and fit	6
How shoe manufacturers influence fit	11
How retailers interpret fit	14
Consumer attitudes toward size and fit	17
Body measurements only factual basis for fit	18
New methods of foot measurement	19
Advantages of new standards	22
Money involved in shoes	24
Conclusions	29
Literature cited	29

Shoe Sizing and Fitting

An analysis of practices and trends

By CAROL WILLIS MOFFETT, collaborator,—Bureau of Home Economics

Toward Better Fitting Footwear

The most important single property of shoes is their fit. If they are uncomfortable or painful to wear, they are not worth the price paid for them, whatever their quality or beauty. They need to fit with accuracy to be comfortable. At present, lack of uniform sizing standards complicates fitting in retail stores. Controversies arise between various branches of the industry as blame for poor fitting footwear is passed from one group to another, because it is difficult to localize the causes. Shoes are produced in this country in a greater variety of length and width combinations than in many countries of the world, which has led some men in the industry to believe that the fit of American shoes needs no further consideration. Each specialized group in the complex chain from raw materials to shoes in use has tended to consider some other group responsible when shoes do not fit.

The purpose of this report is to explore present sizing and fitting practices against the background in which they originated, and indicate new methods through which better sizing and fitting may be developed to the benefit of all concerned.

Mass production poses questions about fit

The present confusion in sizing and fitting practices may result from the fact that answers have not been found to important questions which have arisen with the mass production and distribution of footwear. Among these are:

Why have present methods of sizing and fitting proved difficult to apply in mass production and distribution?

How can a foot be measured so that its contour as well as its dimensions can be recorded by methods which will give the same results wherever they are used and leave as little as possible to individual interpretation? In other words, can the measurement of feet be made a science rather than an art, as it is largely today?

What are the dimensions and shapes of feet in our country?

With more scientific methods of measurement, adapted to statistical analysis of a large amount of data, is it not possible to discover the combinations of specific measurements that best predict the other dimensions of feet?

If a large and representative sample of feet could be measured, would not the shoe industry have knowledge not now available about the predominant types of feet and their distribution in our population?

Would it not be possible to develop uniform standards for sizing and fitting shoes based on exact knowledge of all the feet to be clothed, which would be advantageous to everyone?

Are the technical difficulties insurmountable in translating data about a large number of feet into the shape and dimensions of lasts, the wood forms over which shoes are made?

Without scientifically determined standards for sizing and fitting, can a mass production and distribution industry supply satisfactory clothing for the feet of millions of persons of all ages and sizes?

Who is to undertake the job of developing the necessary methods of foot measurement, take the measurements of a large representative sample of feet, analyze and correlate these, and develop improved standards for sizing and fitting based on information thus gained?

New methods provide answers

Not only have these questions not been answered, but few of them have been considered broadly by many members of the industry. For generations, shoes have been sized and fitted largely on the basis of personal experience. Very few efforts have been made to gather extensive data and interpret them for sizing and fitting shoes. No extensive measurement of the feet of all ages, races, and conditions of people in this country ever has been attempted on the scale the problem merits, partly because methods adapted to such a study have not been available and partly because the usefulness of such a study has not been widely recognized.

Public opinion polls and market research surveys have proved that the intangible thoughts and feelings of people toward particular products or public questions can be measured accurately. The analysis of the data obtained from interviews enables opinions to be understood and classified. It gives manufacturers guidance in marketing products and indicates attitudes toward matters of general concern. Measurement of so concrete an object as the human body is less difficult and the results are just as useful.

However only two studies of body dimensions have thus far been made with a view to aiding clothing industries and consumers (*36, 37*).[1] The Bureau of Home Economics recently measured nearly 150,000 children and 15,000 women in different sections of the United States. Due to limitation of funds and the large number of measurements involved, it was impossible to include those needed for gloves, shoes, or hats. This would have necessitated the taking at one time of so many measurements on each person that he would have been fatigued, and the necessary volunteers could not have been secured.

The data on body measurements revealed that they could be classified according to the physical characteristics which must be taken into account in sizing and fitting clothing. New standards for garments and patterns are now being developed with knowledge of the shape and dimensions of children's bodies, the size intervals necessary to provide good fit for the majority, and the numbers coming within each classification and size (see p. 19). The experience gained in this study offers the opportunity for a fresh evaluation of present methods of sizing and fitting shoes, and points the way to a new factual approach.

---

[1] Italic numbers in parentheses refer to Literature Cited, p. 29.

Background of Present Practices

Industries that owe their existence to advances in science and technology, such as radio, automobile, and mechanical household equipment, originated in an age of mass production and distribution. Their rapid, universal distribution is due in large degree to standardization in tools and methods. Little resistance arose to standardization because these industries were not impeded by craft traditions. Standard sizes for parts were recognized as indispensable to progress.

Craft traditions cling

When the present age of mass production dawned, the making of footwear had behind it 60 or 70 centuries of tradition as an art and craft. The ancient Egyptians and Greeks fashioned sandals from leather with tools not unlike those used by the hand shoemaker today (*25, p. 254*). The introduction of machine technology, when sewing and pegging machines came into use after 1850, was the first major change in ancient methods of shoemaking. It could not be expected to revolutionize immediately all the practices that were centuries old. Within the experience of men still participating in the shoe industry, the processes of making shoes followed the traditions of the individual craftsman.

The early sandal type of footwear did not raise important questions of fit because it was merely a sole held to the foot with thongs. When boots and shoes came into use, they were constructed to order for individual feet, and their fitting properties depended upon the skill of the shoemaker. As late as 1856, one consumer in this country wrote a book which indicated that fit was less important to her than economy. "A Lady," as the author designated herself, advised women in Every Lady Her Own Shoemaker; or, a Complete Self-Instructor in the Art of Making Gaiters and Shoes in these words (*1*):

> Every lady ought to know how to make every article of clothing that she wears * * * So of shoes: some ladies wear out a number of pairs in a year; they will need a new pair for the street every four or six weeks, and when suitable shoes cost from one to two dollars, and have to be renewed so often, it comes to quite an amount in a year.
>
> The art is so simple that it may be learned by any person of ordinary capacity; and it is not so laborious to make a cloth shoe, but that any lady of tolerable health, can make the whole of one without experiencing any injury.

The art did not seem so simple to the men who were engaged in making footwear to individual measurements and getting complaints from their customers when the shoes did not fit comfortably. When 17-year-old John Hanan convinced his father in 1866 that the output of their small factory should be stamped with their name and advertised nationally (*8, p. 29*), it was a new idea that shoes could be produced in one part of the country which might fit unknown persons living far away. The variability of feet had long been recognized, and standards for sizing and fitting seemed impossible. Even today men engaged in the mass production of shoes, making thousands of pairs each season over the same lasts, insist that uniform standards of sizing and fitting are impractical because no two pairs of feet are exactly alike.

After Philip Kirtland began making shoes at Lynn, Mass., in 1635, shoes continued to be made largely for known individuals in nearby communities (*23, p. 11*). The only advantage of the early factories was increase of output by the specialization of craftsmen on a single process. Not until the Civil War, when the demand for soldiers' shoes was too great for the handcrafters to meet, did problems of fit for a mass distribution market become acute. Mass production, in fact, was forced upon a craft industry by that nineteenth century national crisis. When Gordon McKay, perfecter of a machine for stitching soles to uppers, introduced the system of leasing machines to shoe manufacturers on a royalty basis in 1861, manufacturers accepted it to increase production of shoes for soldiers. Resistance to the adoption of machines began slowly to wane (*43, p. 41*). The use of shoemaking machines led to gradual improvement in methods and materials. As output increased more lasts were needed. The production of hundreds of identical lasts became possible with the development of a mechanical last-turning lathe.

While skill and ingenuity were being applied to the invention and improvement of machine processes of manufacture, relatively much less attention was being devoted to adapting craft methods of foot measurement and shoe fitting to the new mass production. The lasts in use before machine manufacture were crude affairs, which the shoemaker padded with pieces of leather to approximate the measurements he took on his individual customer's feet. When last making became a separate industry, measurements were specified by the manufacturer. These last measurements were derived from those of the custom shoemakers, and each manufacturer considered his own set of measurements his trade secret.

Size standards date back to 1886

The lack of uniformity in sizing shoes led to so much confusion that the first action taken by the retailers, when they organized a national trade association in 1886, was to insist that uniform, national standards be adopted for sizing (*2*). After adopting a constitution, they passed the following resolution:

Whereas it is universally conceded by the retail boot and shoe dealers of this country that the adoption by manufacturers of a uniform standard measurement, whereby the goods of all manufacturers of ostensibly the same size and measurement shall be in reality alike would be one of the most beneficial reforms that could be desired in the trade, reducing to a minimum the difficulties of the dealer in selecting his goods on the basis of fitting qualities, and overcoming the present necessity of carrying so great a variety of different makes to insure goods to fit the feet of all customers, thereby enabling him to conduct his business with a much smaller continual investment, reducing in proportion his percentage of profit on his investment, without in any way working injury to the manufacturer; therefore, be it

Resolved, That this convention does unanimously recommend the adoption of a uniform standard of measurements by all manufacturers, and that a stamp showing the ball, waist and instep measurement of each shoe shall be placed thereon, as a guarantee that the requirements of the resolution are complied with. And that upon the promulgation of a system that shall be practical in all respects, each member of this body does hereby pledge himself to not only recommend such standard, but to patronize in preference such manufacturers as will adopt it.

Resolved, That to thoroughly inaugurate this report, the system to be adopted and recommended shall be thoroughly investigated and tested before being promulgated, and a committee of three or five be appointed to receive such scales or systems as are or may be presented, to examine them thoroughly, and to confer with well known and reliable boot and shoe and last manufacturers, said Committee to have full power after such consultation to decide upon the most practical and comprehensive system, and to publish the same in full in the organs of the boot and shoe trade of the country, and to send printed forms thereof to all boot and shoe manufacturers, jobbers and dealers in the United States (*3, p. 197*).

A committee of four retailers was chosen to recommend the size standards, which were published in October 1886. Although manufacturers protested that shoes sold in the South at that time needed to be two-eighths of an inch larger around the instep than those for other parts of the country, and two-eighths larger around the ball for the Middlewest, the standards as adopted included only one set of uniform measurements. Manufacturers who deviated from them were urged to stamp the actual measurements used on the shoes. These standards have remained the base line from which last measurements and shoe sizes still are derived. There is now so much unpredictable variation from them that conditions in sizing and fitting again approach the confusion that existed before 1886.

No research went into size standards   The old standards were based on methods of taking foot measurements used originally by the custom bootmakers. Mass production of shoes was in its early stages of development. Industrial research also was in its infancy. It was natural that sizing and fitting then should be derived from tradition and experience, rather than a search for new methods which might be better adapted to the mass market the industry was beginning to serve. The shoe industry still is less interested in scientific research than industries of similar importance which owe their existence to technological advances. The individual shoe manufacturer has not felt the need to develop an interest in research, nor any national clearing house to stimulate and correlate a search for new facts.

The shoe-machinery companies have taken the leadership in developing mechanical equipment for the manufacturer (*4, 4a*). They will lay out his factory, help train machine operators, and lease or sell machines to him. The great staple material for shoes is leather, and the production of leather still is on the whole an industry separate from the manufacture of shoes. The tanners maintain a research laboratory and some employ scientifically trained leather chemists who have a professional association through which they disseminate reports on their research and experience. Rubber and textile chemists, physicists, metallurgists, and engineers are also associated with the shoe industry. But these are scientists concerned with the materials that go into shoes, not with the way the finished product fits. The individual shoe manufacturer usually buys the materials for his product in a finished or semifinished state. He is largely an assembler of materials that others devise, produce, and improve, on leased or purchased machines which still others invent and maintain.

In a lecture, H. Bradley, director of The British Boot, Shoe and Allied Trades Research Association (which has no counterpart in this country), effectively summarized for an English scientific group many of the factors that have conditioned the attitudes of the shoe industry toward research. In explaining his work since 1924 to develop more scientific methods for taking foot measurements and applying them to lasts in order to improve sizing and fitting practices, he said (*16, pp. 4–6*):

> I would ask you to reflect on the great antiquity of this craft and see how young, in comparison, the industry as we know it to-day really is. Craft tradition is still a very powerful influence.
>
> * * * * * * *
>
> we are watching in the shoe industry the course of development which is natural to all old-time handicrafts. The first stage is the invention of machine tools with which to harness power and displace the hand operation. Consequent upon this is the factory system, requiring growth and improvements in methods of organization both of production and finance. But there is probably an asymptotic limit to the pursuit of development in this direction alone; invention cannot keep up indefinitely its initial rush to displace hand operations by machine tools. Then comes the need for the intense application of the scientific method whereby qualitative personal judgments are replaced by quantitative data which makes possible mass production, standardisation and control, with a steadily rising level of service to the community at large. It is to this stage the shoe industry has advanced—though I would not wish to imply that the industry as a whole recognises the fact.

Until Bradley began publishing reports of research on new methods of foot measurement, no national association in the shoe industry had openly questioned the value of relying almost solely on traditional practices and experiences in sizing and fitting shoes, or had sought for new methods which might be better adapted to a mass production era. Recognition of the need for and possibility of finding better methods is a requisite of progress. In recent years, some individual shoe manufacturers in this country have made earnest efforts to improve the fit of the shoes they produce. This trend may crystallize in recognition that the task is a large one, involving many groups, all of whom will benefit from cooperative efforts.

How Last Manufacturers Arrive at Size and Fit

The last manufacturer is the first in a chain of interests involved in responsibility for the size and fit of shoes, because fitting properties are determined first by the dimensions and shape of lasts. Out of roughly turned hard maple blocks, with moisture content reduced to approximately 6 percent through a long period of seasoning, he fashions the wood forms that give shoes their initial dimensions and shape.

Last making combines art and machines

First, an original model is made by a modelmaker, one of the most highly paid craftsmen in the shoe industry. The most successful have a sculptor's eye for line and form as well as a creative style sense. From this original model, a complete range of sizes is produced on the last-turning machine, which can be adjusted to increase or decrease from the model in all dimensions, and turn lasts for both right and left feet.

Technical skill and care are required to avoid distortion from the original lines of the model in grading sizes. While some manufacturers grade from 9 models (3 lengths and 3 widths), careful ones may make as many as 40 sizes from the original model by checking and adjusting them for uniform size gradations. From these 40 they grade up and down only a half size, while several sizes are turned from 1 model when only 9 are used. The use of many models is more important in women's shoes, where heel heights increase the chances of distortion. A complete size range in women's lasts includes over 130 length and width combinations; whereas that for men is over 100. The United States Army regularly uses 90 different last sizes; others have to be ordered outside this range.

Deriving fit is difficult today The dimensions and shape of the original model are important to the fitting properties of lasts. Art and experience are the main guides today in deriving these from foot measurements. The way feet are measured, the kind of data recorded, and the application of these data to lasts are the outgrowth of craft traditions. Although methods have been refined, they follow essentially the same procedures as those of custom bootmakers. Measurements of feet are taken mainly of length and certain girths. The curves have not been considered susceptible of exact recording. Like the sculptor, a successful lastmaker has an eye and a memory for curves and lines.

The only publication reviewing in some detail the various methods used for sizing and fitting lasts is that edited by F. Y. Golding (*24*). It is an eight-volume technical discussion of shoe manufacturing and merchandising, published in England in 1935. Although American-made shoes sell readily in England and have a reputation for fitting better than the English product, students of shoemaking there have written far more about the technical phases of their industry than have authors in this country. British scientists have published several reports of their research on foot measurement as a basis for sizing lasts (*12*, *13*, *14*, *15*, *17*, *27*, *28*, *29*), though as Bradley indicated in the quoted section of his lecture, they have not found complete appreciation of their work.

A review of the development of foot-measurement methods makes clearer their status today. According to Golding, the earliest reference to measuring feet is in Holme's Academie of Armorie, published in 1688 "where a gauge or shoe measure was used to obtain the length of a foot (*24*, v. *8*, p. *217*). The shoemaker used his hand to take girth measurements, and his apprentice had to carry home in his eye the girth of the master's fingers. A strip of paper was used around 1840 to obtain the size of a foot, "bringing this measure from the lower part of the back of the heel around each side of the foot to the tip of the great toe." Men in the shoe industry today remember having their feet measured in this way.

The tape measure and the size stick gradually came into use and they continue to be the staple tools for sizing and fitting lasts. The length of the foot and the width of certain points can be taken with the size stick; but two persons are likely to pull a tape measure with varying degrees of tension. Only when the same person does all the measuring can comparable results be obtained with this method. Moreover, the tape can record only circumferences, not shape.

Men with many years experience in the shoe industry acknowledge that no scientific method has been developed for recording all the important characteristics of feet that must be taken into account in making lasts. Golding wrote (*24, v. 8, p. 195*), "We loosely talk of the 'average foot,' but who can define the *shape* of the so-called average foot?" He points out (*24, v. 8, p. 169*):

> The size-stick and tape-measure are still extensively used for *measuring* the foot, and their shortcomings are met by memorizing features that cannot be given by the ordinary use of the size stick and tape. Measuring is still too much of the "art" and too little of the science.
>
> The test of systems of obtaining *data* of the foot should be: "Can the same results be obtained by others using the method, and can the data obtained be interpreted uniformly by those who are called upon to use them?"

Anthropometry, the science of measuring the living body, is a relatively new branch of anthropology. And even in anthropometry so far the emphasis has been less on the measurement of feet than on other parts of the body. Bradley notes the paucity of such anthropometrical data. The international agreement among anthropometrists on methods of body measurement, published in 1919, indicates that tracings of the plantar outline and the profile were accepted as standards at that time (*21*). Methods contributed by anthropometry, however, can be utilized to develop a more scientific approach to foot measurement.

Not only are foot measurements for sizing lasts, as now obtained, incomplete and variable from one operator to another, but translation of them into the wood of a last presents difficult problems and opportunities for further variables. Shoes made over a plaster cast of a foot are both ugly and uncomfortable to wear, a number of authorities point out. The foot is compressible and in some portions can be corseted comfortably by a shoe. The problem is to determine just where this pressure may be applied to enhance appearance and comfort.

The Orthopaedic Research Laboratory, maintained in Boston to provide more satisfactory shoes for badly crippled and distorted feet, has conducted research in which incidental findings indicate the degree to which various portions of the foot can be safely and comfortably compressed. In the course of its 15 years of work on these problems, discoveries have been made concerning changes in feet with weight on and off which can be applied also to lasts in general use, though its findings have not been published. Because of the nature of its service, this laboratory has necessarily worked from casts of feet. It has perfected a method for making casts with weight on the foot, and for "shrinking" the casts to the desired proportions for a last. So accurate has this method proved that failure to produce shoes with satisfactory fitting properties is now reduced to less than 2 percent. The cost of this method and the time it requires preclude its use in recording the characteristics of a large, representative sample of feet. But the findings do indicate that foot data could be applied with scientific accuracy to lasts. Bradley also has demonstrated that it is possible to construct instruments which supplement with scientific procedures the intangible artistic sense which must now be relied upon so largely (*14*).

At present, two lasts containing the same volume of wood may have very different fitting properties. Any number of variations may occur which vitally affect the fit and comfort of the finished shoe. The most striking illustration of the relation between volume of wood and its distribution is the fact that lasts of six sizes may contain practically the same volume.

What are "model" feet?
After a last is developed which appears satisfactory, shoes are constructed over it to be tried on "model" feet. In every last- and shoe-manufacturing center, there are one or more pairs of feet regarded as the "most perfect models." These are in demand to try out new lasts. A size 4B for women's shoes is the standard model on which samples first are made. The use of a 4B is so old a tradition in the industry that no one can recall its origin. The traditional use of 7B as a standard model for men's shoes does not seem to represent quite so tenacious a tradition, and some last manufacturers depart from it in favor of 8C or 9C, which are more widely sold. Careful manufacturers usually test models also in larger sizes and narrower widths to satisfy themselves about the potential fitting properties of a particular style of last.

The measurements used in selecting model feet are not entirely uniform. Different communities and manufacturers have their own ideas about the dimensions of a perfect model foot. Because no extensive study has been made of the foot characteristics of the population to be served, experience must be relied upon to select what the manufacturers hope is a typical foot. All of them find it difficult to predict in advance whether a particular style of shoe made over a particular last will prove to be a "good" or an "easy fitter," and hence sell well. An English student of the subject reported (*29, p. 15*):

> When a new last model is being considered, important questions arise, of course, in regard to its shape and general conformation. These questions may have been discussed and settled in the model's favor * * * But an equally important question still remains unanswered after the foregoing has been done. How many feet will the last suit; what proportion of feet in the market for which the last is intended have characteristics corresponding with those of the last? A model may be perfect as regards the interrelations of its different features and shoes made on it may be excellent fitters on feet for which it is intended, but if the proportion of such feet is small the model will never be what the trade knows as a "good fitter" in the sense that it is also a "good seller."

Even if a last model has satisfied the testers, and replicas of it are turned out with care and without distortion, the fitting properties of shoes made over it may be affected by a number of things other than its suitability for many types of feet (see pp. 11, 14, 17).

Size standards do not exist in practice
In a mass-production era, some way had to be found to communicate to shoe manufacturers, retailers, and consumers the size of foot that a given last was supposed to fit. The size standards adopted in 1886 were designed to meet this need. Through the years, however, individuals have been departing from these standards. Men in the industry point out that the departures have produced better fitting shoes, but some of them also are aware that they have produced chaotic conditions in sizing. No true size standards can be said to exist today, because the word "standard" implies agreement on a uniform way of defining and stating size and other character-

istics of a product. Much of the confusion in fitting shoes stems from this disagreement on dimensions for a given size. Several last manufacturers have published "American standards," but these vary in details of measurements.

Length varies with width and toe shape
Sizes are expressed in fractions of inches for both length and width. According to The Shoe and Leather Lexicon (*10, p. 65*), numbering for length begins with 0, which is 4 inches. Each full size above 0 is $\frac{1}{3}$ of an inch longer than the preceding size. These run up to $13\frac{1}{2}$, and then begin again at 1. Thus a size 12 in the first range is 8 inches, and in the second range it is $12\frac{1}{3}$ inches long. The half sizes vary by $\frac{1}{6}$ of an inch.

In practice, lengths are not held strictly to these measurements. According to them, a 4B woman's foot would be $9\frac{2}{3}$ inches in length; feet of models now selected measure either $8\frac{5}{8}$ or $8\frac{7}{8}$ inches. A broad shoe is made longer than a narrow one of the same size, in order to give it a more streamlined appearance. Thus a 7A will be shorter by one-sixth of an inch or more than a 7E. The amount of extension given the toe of the last over the length of the foot is a matter dictated by style or determined by the model-maker. Men's and children's shoes now are given a safe margin of toe extension for the most part, but women's shoe styles today demand a short vamp and efforts are made to keep the extension as short as the type of toe will permit.

Sizes are expressed in over-all length, but most last manufacturers have their own standards for the heel-to-ball length. This usually is held constant, whatever is done with the toe. The additional length given to broader-toed shoes is mostly in the forepart.

Width measurements refer to girths
Widths stated in the letters A, B, C, and so forth, are thought by most consumers to be the width at the sole of the foot. In fact, they represent girth measurements at the ball, waist, and instep. The difference in girth between widths usually is one-fourth inch (*11*).

These also are varied in practice. Individual last manufacturers may add a sixteenth of an inch or more to one girth and take it from another, in their efforts to produce a better fitting last. Furthermore, girth measurements of two lasts may be the same, but the width of the sole, on which the foot treads, may vary.

The width of the heel seat, and the curves at the heel and other parts of the last also are matters on which opinion differs. While experimental changes in last dimensions have undoubtedly produced a wider variety of shapes that enable more feet to be fitted comfortably, shoe sizes have no exact meaning today, because they do not reveal important variations in dimensions and contour.

The classification of sizes for different age groups introduces another variable. This does not present a problem in shoes for adults, but is important for growing children who need shoes adapted to their development as well as to the size of their feet. The National Shoe Styles Conference reclassified sizes for different age groups in 1934 and designated sizes $8\frac{1}{2}$ to 12 as children's shoes (*11, p. 61*). Recently a manufacturer of children's footwear announced that he was reclassifying sizes to "guarantee good fitting." The reason he gave was that a child with a foot requiring a 12½A shoe would be given a misses' style, which was not suitable for her age or activities (7). Children of the same age vary greatly in size; more knowledge is needed of the measurements of children's feet in the various age and interest groups before a uniform classification of sizes for them is likely to prove acceptable.

Footwear styles which originate in fashion trends are interpreted in the shape of lasts, and may be suggested by last manufacturers, stylists, shoe manufacturers, or retailers. As with all clothing industries, fashion changes create continuous hazards. A last manufacturer may spend time and money developing many lasts for each one on which he finally obtains orders. He does not produce lasts in volume in advance of orders from shoe manufacturers.

Even after he has developed a last in the fitting properties of which he has confidence, he must change its dimensions at the shoe manufacturer's pleasure. Some of these changes have to do with style features, but many are based on ideas about what is good for feet. Style is and always will be a matter of personal taste, not susceptible to any scientific measurement except that provided through market research techniques. What is good for feet could be determined by careful research. More scientific knowledge about the characteristics of feet and the relation of shoes to foot health would enable last manufacturers to evaluate for themselves and their customers the soundness of their own and others' ideas about dimensions and contours.

Some individual last manufacturers have shown increasing interest in developing company standards for grading sizes. Recognition by a company of the value of uniform practices based on ascertainable facts usually is a first step toward the development of national standards. These often go through a stage of being considered trade secrets, as they are now. Then comes recognition of the advantages to all of uniform national standards.

How Shoe Manufacturers Influence Fit

Variation exists in the degree of care with which shoe manufacturers select lasts and test them, adapt patterns for uppers to lasts, and mold the materials around the last to make the finished shoe. Those seriously interested in producing better fitting shoes require last manufacturers to check for uniformity the curves of all lasts they order. They also expend considerable time and effort in making samples for trials on model feet of several sizes. The most perfectly designed last may be used, however, to produce shoes that are uncomfortable or inaccurately sized.

Factors other than last shape affect fit
The shoe manufacturer procures patterns for the uppers of shoes from patternmakers. Patterns for soles are provided by the last manufacturers. These have to be graded for the different size combinations as do lasts. A set of pattern dies for cutting materials in all length and width combinations is expensive. The use of old patterns on new lasts which they do not fit accurately

is said to be a common cause of ill-fitting footwear. Patterns need to be carefully adapted to particular lasts, and each style presents its own problems. If the pattern on a pump, for example, is too high at the vamp, the pump will cut into the foot at this point. If the vamp line of an oxford is too low, that shoe likewise will be uncomfortable.

Other difficulties also are cited. Sometimes manufacturers attempt to economize on the number of upper patterns by using the same one for two or more sizes of lasts, to the detriment of fit. There are shoes on the market, according to one investigator, which are marked with different sizes although they are made over the same size lasts.

The fit of the counter used to stiffen the back of the heel, the kind of materials used in the shoe, and the amount of tension applied to them in pulling them over the last, all affect the fit of a shoe. If the counter does not follow the contour or curve of the last heel, the shoe will not fit snugly at this point. If heavy material is used, the last must be more roomy over the ball. Elastic materials or even leather may be pulled too tightly over the last. Shoes, particularly the lower priced ones, are hurried off the lasts since new drying processes have been introduced; overstretched materials are likely to shrink when the last is removed. The size of lasts also may change slightly unless they are protected from variations in the amount of moisture in the air.

Manufacturers' concern with fit is complicated by the blame shoes receive for the prevalence of foot disabilities today *(34)*. Shoes are considered not only as an article of clothing which by their nature may distort feet if they are not properly fitted, but also as a therapeutic device. Crippled or distorted feet may need shoes designed especially for them, but evidence now available indicates that mass-produced shoes cannot fulfill the therapeutic claims made for them.

Studies and estimates indicate that about 70 percent of the population has mild or serious trouble with their feet *(18; 32, p. 100)*. Deformities and blemishes, such as distorted toes, bunions, corns, callouses, ingrowing and thickened nails, lumps over the heel tendon, and shortened calf muscles often are attributable to continuous wearing of shoes that were too tight, too short, or too high-heeled. These disorders, which sometimes cause great discomfort, often could be prevented if individuals always wore shoes that fit their feet properly.

Who's to blame for aching feet? Arch disorders, from which many people suffer, are not caused nor cured by shoes. Recent research has offered evidence that these often are due to maldistribution of weight stresses in the bony framework of the feet. The structure of such feet may gradually collapse with active use and age *(31)*. The causes contributing to pain in the feet may be obscure. They require highly competent individual diagnosis by thoroughly trained medical men who are seriously interested in foot disabilities. The health of an individual is jeopardized by superficial attention to painful symptoms in the feet without searching for the causes of these symptoms. Feet are part of the body and their condition is directly influenced by systemic disorders as well as by their weight-bearing function.

Because of the indifference of the medical profession in general to common foot disorders, manufacturers are led into competitive efforts to treat disordered feet with shoes. No one ever dies of painful feet. So even though earning capacity may be critically handicapped, little serious attention is given by medical men to discovering and treating causes of foot disorders. Other groups with less training than the medical profession requires have taken on concern about foot health. The prevalence of foot disorders and their neglect by the medical profession has encouraged shoe manufacturers, among others, to attempt treatment.

Shoemakers began to take an interest in foot comfort about 1785, when they introduced shoes designed for right and left feet (*25, p. 256*). Before that time a shoe could be worn on either foot and little attention was paid to the anatomy being clothed. When shoes began to be blamed indiscriminately for foot troubles, some manufacturers found it good business to "play doctor" and offer panaceas to the public. Many medical men, although discouraged by their professional journals from believing that mass-produced footwear could prevent or correct individual arch disorders (*5, 9*), have found manufacturers willing to incorporate their theories into shoes. These theories, accepted mainly because of their sales possibilities, represent confused and conflicting ideas. About 450 trade-marked "orthopedic" and "corrective" shoes have been on the market recently. Nearly 200 of these assumed the title "doctor" (*20, p. 205*), until Federal Trade Commission rulings reduced the number.

Laws in every State prohibit laymen from attempting to diagnose and treat the physical disabilities of another individual. A business enterprise organized for profit is recognized as unable either by training or interests to render objective, unbiased professional aid to persons suffering from physical disorders. Some men in the shoe industry are beginning to realize this and to desire a center of competent and authoritative medical guidance on the relation of shoes to foot health, with the medical profession accepting its responsibility for serious attention to common foot disorders. Until this is achieved, individual doctors probably will continue to sell their theories to individual manufacturers, and handicapping foot disabilities will continue to receive less individual scientific diagnosis and treatment than they deserve.

Military men long have recognized that feet must be clothed comfortably if an individual is to be able to concentrate on the job at hand (*33*). When feet hurt, the individual consciously or unconsciously curtails his activities or tries to carry on with a handicap, the seriousness of which is in direct proportion to the discomfort. The interest now shown by shoe manufacturers in foot health may be concentrated to the profit and benefit of all concerned on the big task of fitting their mass market comfortably, which is a responsibility they share with last manufacturers and retailers.

Style often stressed more than fit
Style changes, particularly in women's shoes, also distract attention from emphasis on fit. The shoe manufacturer, like the last manufacturer, works from samples. A single manufacturer may make hundreds of sample shoes from which retailers order, or only a few. The manufacturer with his own retail outlets

must make independent decisions about the styles he offers. Those who compete for the business of independent outlets do little manufacturing in advance of orders except on staple styles. The success of a style depends on consumer acceptance, which is largely unpredictable.

Styles originated in expensive shoes spread rapidly to lower priced footwear, and may as quickly lose their appeal. Retailers who place large orders before a new style has consumer acceptance may be unable to sell them. Unsold stock sometimes is returned to the manufacturer with "don't fit" given as the reason. If a delay occurs between placing an order with a manufacturer and delivery of the shoes to a retailer, which is said to be not uncommon, the retailer's peak buying season may have come and gone. A story is told of a retailer whose order from a manufacturer was delayed 3 weeks at the height of the spring buying season. When the manufacturer finally telegraphed that the shoes were ready and inquired how they should be shipped, the retailer replied, "Ship them in a casket. They will be dead when they get here anyway."

Style hazards harass an industry already highly competitive, particularly in the cheaper shoes in which the bulk of the sales are made. Because shoes are a complicated product whose quality differences are not easily discernible to the ultimate consumers, competition is mainly on price and appearance. In general, prices do not react flexibly to increases or decreases in the cost of materials or labor. Shoes are made to sell within a certain retail price range, and manufacturers specialize within these ranges (*39*). The mortality rate among shoe manufacturing establishments is high. One student maintains that from 1926 through 1934 it averaged 16 percent and usually is not lower than 10 percent a year (*19*). If a manufacturer guesses wrong on styles or makes an obviously poor quality in the price line he has chosen, he is likely to be without business.

Cooperative efforts among several groups that compose the shoe industry are directed toward controlling some of the chaos attendant upon rapid style changes. Basic colors are being selected each season and recommended as style leaders. Common difficulties in sizing and fitting shoes are susceptible to similar cooperative action. There is no agreement now between groups as to what these difficulties are or how they can be mitigated.

How Retailers Interpret Fit

Some retail shoe salesmen are said to have the same kind of intangible sense for size and form that characterizes a good last model-maker. These men are known as "good fitters" in the trade. Others must rely on what they can learn about the dimensions of the shoes in stock and the relation of these dimensions to the instruments they use in measuring customers' feet. While last manufacturers take the first responsibility for fit, and shoe manufacturers can influence it markedly by the degree of attention they give to fitting properties, retailers may defeat the best efforts by selling the wrong shoes to an individual customer.

SHOE SIZING AND FITTING

In selecting his stock, the retailer's attention is necessarily on the styles and colors he thinks will sell best. Many also are concerned with "corrective" shoes for foot sufferers. The retailer is confronted with a large investment in stock if he tries to carry full size ranges for different styles, colors, and classes of shoes. He may stock comparatively few sizes in one style line only to find that it is in demand. This gamble causes some retailers to stock incomplete size ranges and then to compromise with fit in order to sell the shoes on their shelves. The Commission method of selling in many stores intensifies the salesperson's effort to make a sale if possible. When the shoes are a little tight, the customer may be told that they will stretch; if a little too large, that the feet will swell or the shoes can be "tightened."

Measuring hampered by lack of standard sizes
Because there are no uniform size standards today, the measurements of customers' feet cannot be applied with certainty to the shoe sizes in stock unless the retailer is familiar with the dimensions of the shoes he sells. Foot measurement instruments are based on one-sixth of an inch difference in length and one-eighth of an inch difference in width between half sizes. Though the size stick is most commonly used, 20 or more varieties have been produced. Some of these measure the length of the customer's foot and the width at the ball; others also indicate the heel-to-ball measurement. There is no agreement among retailers on whether measurements should be taken with weight on or off the foot, though this may make considerable difference in the size the foot measures.

Shoe salesmen are told that one to three sizes over the customer's foot measurement should be allowed in fitting, depending on the type of shoes (*11, p. 61–62*). At the same time they are informed that shoes marked with the same size are variable in length. A retailer particularly interested in careful fitting of children's shoes reports that "according to the needed width, one must judge the needed length. A 7 is not a 7 in all widths and all lasts, and few fitters have that information" (*22*).

Blind sizes confuse customer and salesman
Sizes often are stamped in shoes by codes that can be interpreted only with the help of a deciphering table. Not all manufacturers and retailers use blind size markings, but many do. Over 160 codes are reported to be in use today to indicate size, and they often resemble stock numbers more nearly than size designations (*11, p. 83–87*). The system of blind marking enables the retailer who does not have a full stock of sizes to sell a size which is not indicated by the measurement of a customer's foot, with the customer none the wiser. It has helped to conceal from consumers the confused practices in sizing and fitting, and also to overcome the resistance of consumers who are more concerned with buying a particular size than with obtaining shoes that fit their feet.

It is possible to put the same adult foot into any one of six sizes, although among these is one which will be a better fit than the others. A man whose foot measures 9B on the size stick may, in addition to that size, be able to put on sizes $7\frac{1}{2}$E, 8D, $8\frac{1}{2}$C, $9\frac{1}{2}$A, or 10AA. The volume of wood in last models of all these sizes is approximately the same in a given style of shoe, but its distribution is very different and they will not all feel the same on the feet. The same is true of women's sizes.[2] In some stores, salesmen must pay for shoes returned as too short, so the tendency is to fit them long and avoid this penalty. The heel will slip in the longer shoe, the ball will not lie over the proper part of the shoe sole, and in other ways the shoe will not fit well.

To check up on retail fitting practices, one last manufacturer selected a girl with a 6½A foot and sent her around to a number of retail stores in different cities to check up on fitting practices. She returned with shoes in a wide range of sizes, but few that he considered a good fit. A shoe manufacturer had a woman buy 12 pairs of shoes at different stores in Washington, D. C. She was satisfied with the fit of all of them, though there were 8 different sizes among the 12 pairs she purchased. Four pairs were judged by the manufacturer to fit her well, and among these were 3 different sizes. More men in the industry are becoming aware that with present sizing practices, fit and size are far from synonymous.

Altering shoes is unsatisfactory
Retailers of shoes selling for $4.95 to $14.95 have reported that it is costing them an average of $100 a month to alter gored shoes or line counters to keep them from slipping at the heel.[3] In shoe-style conferences they state that many shoes do not fit well, particularly at the heel, and that the cost of trying to alter them is becoming too burdensome to be borne. Last and shoe manufacturers claim that many retailers are doing a poor fitting job. So the blame is passed from one group to another for the fact that many consumers today are not getting shoes that fit them properly.

Consumers sometimes are told that alterations will be made which are impossible. For example, if shoes are too short they cannot be stretched. If they are stretched because they are too tight, the seams rather than the materials usually are pulled. Shoes cannot be "tightened" except by putting a lining in the counter. They must be properly proportioned when they are made and properly fitted when they are sold, if the consumer is to be assured of satisfaction, but the temptation to make a sale when the customer likes a particular style is great.

Retailers never have found an acceptable training course for salesmen in shoe fitting. A number of retailers are greatly concerned about the damage poorly fitted shoes can do to feet, and have ceaselessly urged more attention to fit. Recently, the National Shoe Retailers Association has appointed a committee to consider this problem and attempt to devise a satisfactory training course for shoe salesmen. Such work cannot be entirely satisfactory until types of feet are classified, shoes provided for each type, and sizing made uniform.

---

[2] PORTER, O. S. LASTS. Lecture delivered at Mass. Inst. Tech., Dec. 4, 1940. 14 pp. [Typewritten.]
[3] NATIONAL SHOE RETAILERS ASSOCIATION. POOR-FITTING SHOES. Nat. Footwear News 6 (8) : 4. 1940. [Mimeographed.]

Consumer Attitudes Toward Size and Fit

The problem of fitting shoes accurately on all types of feet is complicated further by consumer vanity, according to men in the industry. They point out that many persons are more interested in selecting shoes that appeal to their eyes than that fit their feet. This usually is evident in wanting shoes too short or too tight, in the hope of making feet appear smaller than they are. Men as well as women are said to be susceptible to this bit of wishful thinking. Small children usually are so delighted to have a new pair of shoes that they are not critical about the way shoes feel on their feet.

Vanity clouds the issue Many consumers hold the erroneous belief that shoes are necessarily uncomfortable when they are new and must be "broken in." This causes them to be less meticulous than they should be about insisting that shoes fit comfortably when they are purchased. Influenced by style promotion that stresses such ideas as "flatter your feet," and hopeful that shoes will become more comfortable as they are worn, many consumers apparently give retailers the impression that appearance is more important than accurate fit.

Because their feet may measure the same on the size stick time after time, consumers think of fit in terms of standardized sizes. Few are aware that shoes in different styles and lasts, though marked with the same size, may differ greatly in the way they feel on the feet. Many insist that a certain size fits them best without realizing the influence of styling on present sizing practices. Partly in an effort to overcome consumer misinformation, as well as for other obvious reasons, people have been encouraged to rely upon shoe salesmen to fit them with the shoes they should have.

There is a striking difference between this attitude toward consumers when trying on shoes and the demands made upon models who try out the samples for shoe and last manufacturers. They rely upon the models to tell them how the shoes feel, and say "a good model's head registers what is going on in the feet when new shoes are tried."

When medical men cannot give foot sufferers permanent relief by discovering and treating the causes of their disorders, consumers often take their troubles to shoe salesmen, who undertake to tell them what they need. They have grown up in an era when shoes have been given most of the blame for "fallen arches," and are influenced by the claims made for "corrective" and "orthopedic" footwear. This confusion of fit with therapy has tended to make consumers dependent on other laymen to decide what is good for their feet. Baffled, and expecting too much from shoes and the free advice given them in shoe stores, these customers wander from store to store seeking relief from foot disabilities.

As the consumer movement has grown in size and momentum, more individuals are trying to arrive at independent judgments about the things they buy by studying these in relation to their needs and incomes. This has created a more active attitude among consumers toward removal of unnecessary obstacles to the satisfactory use of their incomes. Many of them are concerned about the difficulties they now have in getting shoes that fit their feet comfortably.

Body Measurements Only Factual Basis for Fit

This review of present difficulties in sizing and fitting shoes would serve little purpose unless methods for improving the situation also were explored. Experience with new methods for sizing children's body garments indicates an approach which is proving successful. Like shoes, children's garments are sized over forms, or manikins. Age was the basis for size, but manufacturers did not have uniform measurements for the various age designations. Garments marked for the same age have varied widely in dimensions. Mothers either have had to take children to stores to try on garments they wished to buy, or have a number of sizes sent home. The net result was great waste of effort for mothers, a high rate of returns to stores, and conflicts between manufacturers and retailers. After a careful review of the situation the Bureau of Home Economics believed that these problems could be solved by measuring a representative sample of the children of the United States and learning what dimensions would serve as the best basis for sizing manikins and patterns, and stating the size of garments.

Ground work laid by children's project

Because no extensive project to measure thousands of children ever had been undertaken, no methods were available which could be used by operators in different sections of the country, with comparable results. The necessary methods and instruments were developed by the Bureau, based on the techniques of anthropometry. When funds for the project became available, men and women who were to take the measurements received training and detailed instructions for carrying out their work.

In 1937, 36 different measurements were taken on 147,088 children of 4 to 17 years of age, in 16 States and the District of Columbia, under the direction of the Bureau. These measurements were correlated and analyzed by statisticians with the following objective (*35, p. 2*):

> Essentially the problem of sizing garments is one of finding that measurement or combination of measurements which best predicts the other dimensions of a child's body. This is especially true when what is required is the creation of a representative form or model upon which standard garments may be manufactured. It follows, therefore, that the best choice is that measurement or combination of measurements which is most closely related to the greatest number of the others, provided, of course, it also satisfies the further criterion of being practicable.

As a result of this study it was found that age was the poorest indicator of a child's size, while a combination of height and girth of hips was both satisfactory and practical as a predictor of other body dimensions. For given girths and heights, it was possible to find the figures for the other 34 measurements most closely associated with them. The numbers of children included in the various combinations of height and hip girths were discovered. Almost half of them came within one classification, which could be termed "regular." About one-fourth were above or below regular, in that their hips were larger or smaller in proportion to height. A very small percentage came outside the dimensions indicated in these classifications. Sizes could be devised according to the types of figures into which the largest groups of children fell. Intervals between sizes, which were found to be unequal, could be arranged to conform to known size progressions. All the detailed measurements, such as arm length, etc., which were associated with given heights and hip girths were known. A new field of specialty manufacture was opened for known variations from prevailing types.

Under the auspices of the American Standards Association, the data from this study were utilized by a committee representing national trade groups and associations of manufacturers, retailers, and consumers of children's clothing. Members of the committee have been working together to arrive at a consensus on new size standards for the body garments of boys and girls between the ages of 4 and 17. As the standards are developed, new manikins must be maufactured according to the dimensions for each size. The process of transition from the old sizing practices to the new size standards necessarily takes time, but manufacturers now can proceed with more exact knowledge of the market they are serving. Many of the difficulties that chaotic sizing of children's garments have created are due to disappear.

New Methods of Foot Measurement

A new approach to sizing and fitting shoes involves many of the same problems encountered in the study of children's body measurements. First is the problem of how the foot is to be measured by a number of operators in various parts of the country to obtain information that is comparable and can be analyzed by statistical procedures. This involves also the question of what measurements will be most revealing in their application to the sizing of shoes. After methods have been developed, the representative sample must be selected, operators must be trained and funds made available for the study. The data obtained must be analyzed and correlated to find the measurements that best predict the size and shape of feet. Representative and informed last technicians, shoe manufacturers, retailers, and consumers could then be drawn together in committees under the auspices of a national standards agency to determine how these data may be applied to sizing and fitting shoes. Experience has shown that unless all the parties at interest can reach a consensus, the efforts of any one group to develop new standards or improve old ones by voluntary methods is not likely to succeed. The makers, sellers, and users of a product each have a contribution to make, and each needs to be satisfied that the proposed standards would be beneficial to all concerned.

Although the human foot is the one constant factor with which the shoe industry must deal, the question of what a foot is never has been answered satisfactorily. It is characteristic of all feet that they change in shape and size from babyhood to adulthood, and may change with added body weight and different ways of use during adulthood. Normally, they are alike in structure because they have evolved to their present form as a result of centuries of standing, walking, and running. No one has difficulty in recognizing a foot, but a precise statement of its size, shape, and rate of growth or change has proved elusive. Such a statement is necessary for making an article of clothing in mass production that is as unalterable to suit individual variations as are shoes.

Typical foot contours needed

If feet varied only in length and width, the problem of measuring them and sizing shoes to fit would be simple. But some arches are low and others high, with variations between these extremes. Insteps vary in the same way. Heels are different in width, and protrude at the back with different curves to which the shoe must be fitted if it is not to cut into the heel tendon or slip at each step. Some feet have short, stubby toes; others have long ones which may be either thick or thin. Toes may be almost straight across their ends, or angled, or the second toe may be longer than the first. The length from heel to ball varies, as do other characteristics. The angle to which the front part of the foot turns inward or outward in relation to the heel is not always the same. Differences exist in the compressibility of feet, and their extensibility under weight bearing. The foot's contour changes with different heel heights.

No study ever has been made of all these characteristics in large numbers of feet. Nor has any statistical analysis been made to determine which characteristics most frequently are combined, or to learn which particular characteristics serve as the best predictors of the others. No knowledge is available about the distribution of particular sets of characteristics among the feet of our own or any other population. Shoemen sometimes classify feet in four types: Long and narrow; short and square; high- or low-arched; soles with inflare or outflare. They acknowledge that several of these characteristics may be combined in the same foot. They point out that feet are more varied in this country than in many others, because of the diverse origins of our people. Careful study and analysis of all important characteristics would make possible a classification of feet according to types, and would reveal the most practical size intervals for fitting them with shoes.

British research helpful

In 1924, The British Boot, Shoe and Allied Trades Research Association began research under the direction of Bradley (*12, 13, 14, 15, 17, 28, 29*), to discover more scientific ways to define a foot and apply statistically analyzed foot data to sizing and fitting shoes. Bradley developed a method of enclosing the plantar outline of the foot in a quadrilateral, which enabled him to state the size of some angles in geometrical terms. On this basis he devised a formula for a more precise definition of some of the characteristics of a foot, and an instrument for recording some angles. He reports that his work received some application to lasts and shoes in England before 1939.

Shoemen and medical authorities in this country who have studied published reports of Bradley's work have been interested in the possibilities opened by his approach to an old and baffling problem. They have raised the question whether he did not start with the second rather than the first step. They suggest that it seems necessary first to find a method for recording accurately each segment of the foot, determining the interrelationship of these segments, and letting the formulas grow out of the predominant and associated characteristics which would thus be discovered. Bradley's method, however, was the first adapted to statistical analysis of data on large numbers of feet.

Exploratory study started

The Bureau of Home Economics now is doing research on new methods of foot measurement, based on the experience obtained in the measurement of children, Bradley's research, and other work. One of Bradley's instruments has been imported for further study. Out of this work is expected to evolve a method that can be used by different operators with comparable results, recording curves as well as linear dimensions, and adapted to statistical analysis of large amounts of data. The advice of representative technicians with experience in adapting foot data to lasts and of medical authorities is being sought to insure that the method will be practicable and sound in its relation to shoe manufacturing and foot health problems.

The measurement of large numbers of feet in the population probably would be a big undertaking. When individual companies or trade associations have attempted limited measurement projects, it has proved difficult for them to obtain cooperation from those whose feet they wanted to measure. The measurement of children by the Bureau of Home Economics was undertaken in response to repeated requests from individual consumers and consumer organizations, as well as companies interested in improving sizing and fitting practices. Some organizations recently have shown the same interest in developing improved standards for sizing and fitting shoes.

Like the proverbial hair shirt, the irritations caused by the confusion in sizing and fitting have been borne patiently for so long that many individuals believe they are inevitable. This attitude usually is characteristic of industries deeply rooted in traditions. Considerable effort is required to shake off the inertia that all traditional practices create, in order to take a fresh look at an old problem, and reflect upon better ways of handling it than have appeared possible in the past.

Cooperation of trade groups essential

Sizing and fitting shoes cuts across the lines of interest of the various groups that compose the shoe industry. Trade associations have developed around the particular interests of each group. All are concerned with the end product—the shoe. Difficulties attendant upon sizing and fitting affect them all, as well as the consumers they serve. In the past there has been no central agency or clearing house for cooperative approach to matters of mutual concern. There is a tendency now among men in the shoe industry to develop the kind of cooperative activity necessary to tackling such complex problems as sizing and fitting.

When problems are neglected until the difficulties surrounding them result in confusion and growing irritations, precipitate action in search of relief often is attempted by one or another of the groups involved. The original size standards for shoes are an illustration; they were forced on reluctant manufacturers by retailers after only 2 months of study. Time was not taken to analyze the needs in fitting a mass market or the best ways to meet them; existing measurements were averaged and promulgated as a standard. If new size standards should be developed on the basis of a large measurement project, stocks of lasts now on hand might not conform to these standards. A period of transition would be necessary to avoid disruption. The standards could be applied to new lasts as they were ordered, and thus all last stocks could be replaced with whatever speed an individual shoe manufacturer might choose.

Men in the shoe industry are acutely conscious that feet, like all human structures, are variable. They know that the two feet even of the same individual may not be exactly alike. This has seemed to many to present insuperable difficulties. Experience has shown, however, that when large amounts of data are collected and analyzed, patterns emerge that are characteristic for large groups. Human bodies cannot be standardized, but their characteristics can be known and classified as to the frequency with which they occur. This kind of information is necessary in mass production to provide garments that fit large numbers of persons. It is particularly important to the shoe industry, whose product must fit with accuracy to be comfortable and is not susceptible to much alteration.

Advantages of New Standards

Science seeks to discover heretofore unknown facts, not by repeating the experience and practices of the past, but by invention of new ways to approach old problems. The shoe industry may benefit by search for and application of scientifically discoverable facts. Mass production is particularly dependent upon discovering unknown factors that interfere with the development of satisfactory products. It must standardize its methods and practices in order to produce thousands of like articles most quickly and economically. Uniform standards for sizing and fitting garments on the basis of known body dimensions and conformation is one of the most important needs of all clothing industries. Heretofore the necessary facts have not been available to accomplish this end.

Foot types and sizes clearly defined
The shoe industry has attempted to develop lasts which take account of the fact that feet might be classified by types. Its traditional methods of measurement, refined though they are today, have not proved entirely satisfactory for discovering exactly what characterizes a particular type or with what frequency a type occurs. If this can be discovered through new methods, production for a mass market will be simplified.

Size ranges now include a large number of sizes for each class of shoes. The number of size combinations needed to fit the majority of our population comfortably cannot be arrived at by guessing. Research is needed on the tolerance in size that is best adapted to proper fitting. We have seen that now one foot can be put into shoes of several sizes, although it is recognized that the fitting properties of all these sizes on a given foot are not the same. If large numbers of feet are measured, some authoritative answer may be found for this question of how many sizes are needed.

Size intervals keyed to majority of feet

Size intervals now used are evenly dispersed, progressing for half sizes by one-sixth of an inch in length and one-fourth of an inch in girths, if the old standards are followed. The practice of using one-third of an inch as the length interval between full sizes seems to have originated when "three barleycorns made an inch," before standards for weights and measures were developed. Studies of anatomy indicate that the body does not grow in such regular progressions, nor does it differ in various portions by evenly dispersed intervals. Measuring each segment of the foot would disclose exactly how and where feet differ in dimensions and conformation. New size standards probably would call for unequal dispersion of size intervals, as has already been concluded from work on foot measurements (*27, pp. 4–5*), and as the study of children's body measurements has indicated. Shoemen now believe that as the size of feet increases, there is progressively less need for the present fine size gradations.

Effective length for lasts

The suggestion has been made by the British Boot, Shoe and Allied Trades Research Association that lasts should always be sized according to their effective length, that is, the length of foot which the last is designed to fit. While this is attempted now, there is variation in practice. If all shoes were sized according to their effective lengths, such waste and confusion about fit as arose with the open-toed styles would not be likely to occur. More data on the foot and particularly on the shape of the forepart would enable effective lengths to be stated accurately.

Location of contours

Men who have worked on fitting state that while arriving at uniform statement of effective length would reduce confusion, the distribution of the wood also is a vital factor in fit. The compressibility of various portions probably can be determined by research. The essential factors in the distribution of the wood also may be revealed by recording and analyzing the dimensions and curves of every portion of the foot. The adaptation of foot measurements to lasts raises important technical problems which members of the industry, who have been trying to improve fit despite all the handicaps under which they labor, give evidence of being able to solve.

Retail instruments adapted to lasts

Experience shows it would be possible to devise foot-measurement instruments for use in retail stores which are adapted to known types of lasts. Classification of feet would make it possible to devise lasts for the various types, and invent fitting instruments adapted to these lasts. These could register automatically the last best adapted to a given type of foot.

Known specialty demand

Judging from the experience with children's measurements, probably 4 or 5 percent of the feet of the population would fall outside the majority classification. The exact shape and size of these feet would be known, however, and a new field of service opened for shoe manufacturers and retailers. Special shoes could be provided by those in the industry who might wish to serve this portion of the population, based on knowledge of their numbers and the characteristics of their feet.

Improved definitions of unfair trade practices
Much of the waste and conflict that now goes with trying to pass the blame from one group to another for shoes that "don't fit," could be eliminated without recourse to lawsuits if the industry had uniform and improved methods for sizing and fitting shoes. With new standards for sizing and fitting, based on accurate knowledge, the industry could cope more intelligently and cooperatively in handling its own unfair trade practices.

The need for standard sizes was recognized early in the history of the shoe industry. These standards are outmoded now, as the departure from them indicates. A standard has no meaning unless it applies to accepted definitions and uniform practices. The standard weight of a pound or the standard length of an inch would no longer be considered standard if each merchant departed from them at his own discretion, or if different companies had varying rules for their own use. Present size designations imply agreement on dimensions that does not exist in fact. There is some recognition that new standards based on a combination of experience and new data would provide a common language in sizing and fitting shoes which would be helpful to all concerned.

Money Involved in Shoes

The economic importance of shoes to the individual consumer's budget and to the Nation is so great that attempts to improve the product deserve serious consideration. Over 131,000,000 persons in this country create a demand for footwear that has made the shoe industry of major importance in our industrial era. Shoes are produced to facilitate sports, minimize industrial hazards, and complement every type of dress, as well as fulfill their ancient function of protection of the feet. Keeping feet shod requires a relatively large proportion of the clothing expenditures of individuals and families.

The study of consumer purchases, conducted in 1936 by the Bureau of Home Economics and the Bureau of Labor Statistics with the cooperation of the National Resources Planning Board, the Work Projects Administration, and the Central Statistical Board, provides data on expenditures for footwear for each family member at different income levels in various parts of the country (*30, 38, 42*).

Shoes bulk large in clothing budget
One-third of the Nation's families and single individuals had incomes of less than $780 during the year 1935–36 (*26*). At the family income level $500–$999 wives who had expenditures for clothing during the year spent 21.5 percent of their clothing funds for footwear in farm communities of the North and West, as compared with 19.5 percent in large and middle-sized cities of New England and the East North Central regions. Husbands, and boys

and girls in the age group 12–15 years used even larger proportions of their clothing outlays—24.3, 27.0, and 25.5 percent, respectively, in the farm unit; 22.3, 29.5, and 32.5 percent in the urban unit cited above. Children from 6 through 11 years of age spent less during the year for their shoes and overshoes than did their older brothers and sisters (from 12 through 15 years), but their expenditures for footwear took a larger share of their clothing money.

The middle third of the Nation—13 million families and single individuals receiving from $780 to $1,450 during the year—spent larger aggregate amounts for their footwear than did the lower third of the population, as would be expected. But this middle income group used a smaller proportion of its total clothing funds for their shoes and overshoes than did the lower third. The upper third, with incomes of $1,450 or over, continued the trend of larger aggregate amounts spent for footwear, though this represented a smaller share of the total clothing outlays. Thus footwear follows the pattern of spending typical for food, that prime necessity of life which takes the largest proportion of the budget, increasing in amount and decreasing in proportion as income rises.

A farm husband at the income level $500–$999 with expenditures of $5.78 for shoes and overshoes (the average outlay of the group) would buy only one pair of work shoes at $2.80 during the year if he followed the pattern of spending of others in his income group (table 1). His street shoes, costing $3.49 a pair, would have to last almost 3 years before they could be replaced. The balance of the footwear money would be needed for arctics and rubbers, for shoe laces and repairs, and for polishes. At the income level $3,000–$4,999, a farm husband would pay $3.22 per pair for his work shoes, instead of $2.80; $4.28 per pair for street shoes, instead of $3.49. His work shoes would be replaced every 9 months; his street shoes, about every 1½ years. Thus at every income level, shoes for husbands, and for other family members as well, would have to be worn for a considerable length of time. The comfort and efficiency of an individual, therefore, would be affected for an appreciable period if shoes purchased did not fit properly.

A study of the clothing expenditures of 209 professional women, made by the Bureau of Home Economics from their clothing records kept during 1939, indicates a wide variation in the amounts of funds used for footwear and the price paid per pair of shoes. The women who spent between $100 and $150 for total clothing purchases during the year used $21 (the group's average) for their footwear, while those who spent between $300 and $350 used almost twice that amount, as is shown by the following figures:

Item:	Level of spending for clothing		
	$100–$149	$200–$249	$300–$349
Number of women	44	37	20
Average expenditures for clothing	$128	$227	$324
Average expenditures for footwear	$21	$28	$40
Average number of pairs of shoes bought	3.3	4.3	5.7
Average price paid per pair of shoes	$5.54	$5.95	$6.61

This study reports:

It was the exceptional woman who bought more than six pairs of shoes or fewer than two pairs during the year. Only 7 percent of the 209 women bought

TABLE 1.—FOOTWEAR EXPENDITURES: Average expenditures for all footwear, percentage of total clothing expenditures allocated to footwear, and average expenditures per person and per pair for specified kinds of shoes, selected income classes and sex-age groups, 4 analysis units, 1935–36 [1]

Analysis unit and family-income class (dollars)	All shoes and overshoes [2]		Average [3] expenditures per person for—			Average [6] expenditures per pair for—			All shoes and overshoes [7]		Average [3] expenditures per person for—		Average [6] expenditures per pair for—	
	Average expenditures [3]	Percentage of total clothing expenditures [4]	Work shoes	Street shoes [5]		Work shoes	Street shoes [5]		Average expenditures [3]	Percentage of total clothing expenditures [4]	Street shoes [5]	Dress shoes	Street shoes [5]	Dress shoes
	Dollars	*Percent*	*Dollars*	*Dollars*		*Dollars*	*Dollars*		*Dollars*	*Percent*	*Dollars*	*Dollars*	*Dollars*	*Dollars*
	Husbands								Wives					
North and West farm unit:														
500–999	5.78	24.3	2.91	1.20		2.80	3.49		5.13	21.5	2.48	1.58	2.74	3.10
1,000–1,499	6.94	22.8	3.33	1.60		2.91	3.72		6.49	20.3	3.03	2.16	3.02	3.36
3,000–4,999	9.29	18.3	4.26	2.73		3.22	4.28		9.10	16.2	4.05	3.35	3.83	4.18
North and West village unit:														
500–999	5.68	22.5	2.21	2.15		3.05	3.67		5.79	21.4	2.83	1.91	3.01	3.17
1,000–1,499	7.74	19.5	2.63	3.47		3.34	4.11		7.83	18.4	3.73	2.66	3.43	3.68
3,000–4,999	14.62	13.8	1.64	8.85		4.59	5.84		15.75	11.4	6.18	6.34	4.73	5.26
North Central and West small-city unit:														
500–999	5.92	21.0	2.19	2.58		3.15	3.79		6.22	21.3	3.02	2.24	2.91	3.42
1,000–1,499	8.15	18.5	2.70	3.66		3.35	4.11		8.02	16.9	3.69	2.87	3.42	3.72
3,000–4,999	13.95	12.4	1.92	8.33		5.14	6.42		16.24	12.0	7.62	6.04	5.40	5.82
New England and East Central—2 large and 5 middle-sized cities:														
500–999	5.14	22.3	1.57	2.60		2.02	3.23		4.88	19.5	2.80	1.16	2.69	2.69
1,000–1,499	7.15	18.3	1.66	3.98		2.86	3.70		7.30	16.9	4.36	1.75	3.19	3.14
3,000–4,999 [9]	13.92	12.6	1.05	9.63		4.30	5.81		14.90	12.1	8.63	4.12	4.99	4.88
	Boys 12–15 years								Girls 12–15 years					
North and West farm unit:														
500–999	6.22	27.0	1.93	2.27		2.30	2.56		5.67	25.5	2.59	1.69	2.29	2.42
1,000–1,499	6.56	25.0	2.28	2.79		2.40	2.66		6.96	24.3	3.46	1.66	2.33	2.50
3,000–4,999	7.42	22.0	2.47	3.00		2.40	2.77		9.68	22.1	4.27	2.97	2.70	2.76
North and West village unit:														
500–999	5.77	28.6	.98	3.30		2.38	2.45		5.39	29.8	2.62	1.69	2.29	2.64
1,000–1,499	7.06	25.6	.97	3.93		2.57	2.63		7.89	25.0	3.37	2.77	2.44	2.76
3,000–4,999	14.52	22.6	1.64	8.20		2.58	4.29		12.93	22.8	4.69	4.34	2.82	3.26

North Central and West small-city unit:											
500–999	5.60	30.3	--------	3.40	2.44	2.50	5.63	29.6	3.81	1.01	2.36	2.02
1,000–1,499	7.14	27.5	.98	4.27	2.60	2.78	7.48	24.3	3.67	2.08	2.49	2.49
3,000–4,999	13.18	21.7	1.17	7.60	3.70	3.98	12.75	17.7	4.98	3.60	3.66	3.77
New England and East Central—2 large and 5 middle-sized cities:											
500–999	4.17	29.5	--------	[11]3.20	([10])	([10])	6.53	32.5	4.51	.82	([10])	([10])
1,000–1,499	6.50	27.1	--------	[11]5.10	([10])	([10])	8.46	27.9	4.67	2.17	([10])	([10])
3,000–4,999[9]	11.22	20.0	--------	[11]8.61	([10])	([10])	14.96	18.2	8.66	3.40	([10])	([10])

[1] Data for farm, village, and small-city units from U. S. Dept. Agr. Misc. Pub. 422 and 428; data for large and middle-sized cities from U. S. Dept. Labor Bul. 648, vol. 3.
[2] Shoes for work; street, sport, or other use (dress, dancing, house or bedroom slippers); rubber or leather boots; arctics; rubbers; shoe polish, shines, and repairs.
[3] Averages are based on the total number of persons in each class having expenditures for clothing, regardless of whether they had expenditures for footwear.
[4] Percentages are based on the total clothing expenditures of persons in each class.
[5] Shoes worn at work by office and store workers were classified as street shoes and were distinguished from those worn for farm work and other similar labor, classified as work shoes. Includes shoes worn by boys at school or play. In U. S. Dept. Labor Bul. 648, work and street shoes for boys aged 12–15 are combined in one group, "school" shoes.
[6] Averages are based on the total number of pairs of shoes purchased.
[7] Shoes for street, sports, dress, or other use; house slippers; artics, galoshes; rubbers; shoe polish, shines, and repairs.
[8] Includes shoes worn for daytime wear on street, in house, or at school, and those worn by girls at play.
[9] Data for this income group were obtained by combining the income classes $3,000–$3,999 and $4,000–$4,999. (See U. S. Dept. Labor Bul. 648.)
[10] Data not available in U. S. Dept. Labor Bul. 648.
[11] Data for work and street shoes combined.

one pair or none; only 12 percent bought more than six pairs. Two-fifths bought three or four pairs * * *

Bills for shoes were almost four times as great at the highest level of spending [$350 or more] as at the lowest [under $100], $38.39 compared with $10.43. The women at the former level not only bought more shoes, an average of 4.6 pairs compared with 2.6, but they also tended to pay more per pair.[4]

Prices paid for shoes varied from 35 cents to $16.17, but the type of shoes also varied from beach sandals to sturdy, well-made street shoes.

To supply consumer demand in 1939, footwear with a value of $784,653,702 (including rubber and rubber-soled boots and shoes) was produced, ranking twelfth in value of product among the 177 leading industries reported by the Bureau of the Census. The retail value of these shoes, if computed with a 40-percent mark-up to take into account sale prices and lower than customary mark-up on cheap footwear, would be $1,307,756,170. With less than 1 percent of shoes remaining unsold this figure indicates that clothing the feet accounts for a relatively large proportion of the total retail sales of $42,041,790,000 reported for 1939 (*41*).

Comparatively little spent on research
The shoe industry increased its output by nearly 120,000,000 pairs in the 10-year period 1930–39. In the latter year, it was providing over 424,136,000 pairs of boots, shoes, and slippers other than rubber footwear. The production in 1939 was the highest ever recorded, making available a per capita supply of 3.23 pairs of which it is estimated that 3.13 pairs per person were sold (*6, p. 122*). The Bureau of Foreign and Domestic Commerce reports that this country has the highest per capita production and consumption in the world, and has been the leader in adapting machine techniques to the manufacture of shoes (*40, p. 30*).

Despite its large volume of business, the shoe industry does not report expenditures on research to agencies that gather this information. A recent Federal Trade Commission study of the financial reports of 16 principal corporations operating in the leather boot and shoe industry indicates that none of them reported any expenditures for research and development (*44*). Men in the industry say that a great deal of experimentation goes on along a trial-and-error pattern, without the carefully controlled procedures that distinguish scientific research from empirical deductions.

Difficulties in sizing and fitting shoes raised by mass production and distribution are not inducing a return to making footwear on order from individuals. About 1,070 manufacturers report to the Bureau of the Census. Approximately 900 manufacturers are said to produce 99 percent of the leather footwear manufactured in this country, but 300 factories make 90 percent of the output (*40, p. 30*). If fitting is to be improved, it obviously must come through search for new methods better adapted to mass production than those used in the past.

---

[4] MONROE, DAY; PENNELL, MARYLAND Y.; and ROSENWALD, RUTH. HOW PROFESSIONAL WOMEN SPEND THEIR MONEY. AN ANALYSIS OF RECORDS FOR 1939. U. S. Bureau of Home Economics. 38 pp., illus. 1940. [Mimeographed.] (See pp. 29–30.)

Conclusions

This exploration of shoe sizing and fitting practices and trends leads to the following conclusions:

1. The making of footwear, one of the most ancient crafts, has become a major mass production and distribution industry in this country within the last 50 years. Machine technology has revolutionized centuries of handcraft traditions. Standardization of tools and methods has developed manufacturing capacity that now exceeds consumer demand. Tradition, however, still influences attitudes toward sizing and fitting shoes.

2. Because the organization of the shoe industry is complex and operates in specialized groups, no concerted effort has been made to tackle fitting problems raised by mass production. Competitive conditions in the industry and the influence of styling have served further to inhibit cooperative action on this common problem.

3. A clothing industry which serves a mass market needs nationally accepted sizing and fitting standards, based on exact knowledge of body dimensions and conformation, and classification of these according to the frequency with which they are found in the population. This is particularly true of shoes, which must fit with accuracy to be comfortable, and are not susceptible to much alteration after they are made. Shoes can damage feet if they do not fit properly and thus impair the efficiency and earning capacity of an individual.

4. Scientific research on new methods of foot measurement adapted to the study of a large, representative sample of the population and capable of producing uniform results that can be analyzed by statistical procedures, is a requisite to developing new size standards.

5. Measurement of large numbers of feet and analysis of the data obtained would have to be undertaken by an agency capable of serving all groups in the industry as well as consumers. Successful development of national standards for sizing and fitting shoes on the basis of these data depends upon the cooperative efforts of all the parties at interest, working together through the procedures of a national standards agency.

Literature Cited

(1) ANONYMOUS.
    1856. EVERY LADY HER OWN SHOEMAKER; OR, A COMPLETE SELF-INSTRUCTOR IN THE ART OF MAKING GAITERS AND SHOES. 39 pp., illus. New York.

(2) ——
    1886. THE RETAILERS AT PHILADELPHIA. Shoe and Leather Rptr. 42: 152–153.

(3) ——
    1886. THE RETAILERS' CONVENTION. Shoe and Leather Rptr. 42: 197.

(4) ——
    1933. "BUT BUSINESS IS ALWAYS GOOD." Fortune 8 (3): 34, 41, 110–113, illus.

(4a) ——
    1933. AND THERE'S A COMPANY CALLED COMPO. Fortune 8 (3): 42–43, 114–115, illus.

(5) ANONYMOUS.
    1939. FOOT TROUBLES—A NEGLECTED FIELD. Arch. Phys. Therapy 20. 239–240.

(6) ———
   1940. TAKE A LOOK AT OUR INDUSTRY'S RECORD. Boot and Shoe Recorder 118 (18): [68–69], 122, 248, illus.
(7) ———
   1941. NEW SIZE RUNS FOR CHILDREN'S SHOES. Boot and Shoe Recorder 118 (25): 45, illus.
(8) ALLEN, FREDERICK J.
   1922. THE SHOE INDUSTRY. 415 pp., illus. New York.
(9) AMERICAN MEDICAL ASSOCIATION.
   1935. ACCEPTANCE OF SHOES. Amer. Med. Assoc. Jour. 104: 1503.
(10) BOOT AND SHOE RECORDER.
   1926. THE SHOE AND LEATHER LEXICON. Ed. 5, rev., 87 pp., illus. Boston.
(11) ———
   1935. THE SHOE AND LEATHER LEXICON. Ed. 9. rev., 87 pp., illus. New York.
(12) BRADLEY, H.
   1924. FOOT MEASUREMENT. Brit. Boot, Shoe and Allied Trades Res. Assoc. Res. Rpt. No. 2, 14 pp., illus. London.
(13) ———
   1924. FOOT MEASUREMENT. Brit. Boot, Shoe and Allied Trades Res. Assoc. Res. Rpt. No. 13, 12 pp., illus. London.
(14) ———
   1928. A FOOT AND LAST MEASURING INSTRUMENT. Brit. Boot, Shoe and Allied Trades Res. Assoc. Res. Rpt. No. 20, 13 pp., illus. London.
(15) ———
   1928. FOOT MEASUREMENT. Brit. Boot, Shoe and Allied Trades Res. Assoc. Res. Rpt. No. 22, 8 pp., illus. London.
(16) ———
   1933. PHYSICS IN THE BOOT AND SHOE INDUSTRY. [London] Inst. Phys., Phys. in Indus. No. 19, 20 pp., illus.
(17) ——— and MCKAY, A. T.
   1929. LAST MEASUREMENT. Brit. Boot, Shoe and Allied Trades Res. Assoc. Res. Rpt. No. 25, 21 pp., illus. London.
(18) COOK, ROBERT J.
   1922. REPORT OF THE ORTHOPAEDIC EXAMINATION OF 1393 FRESHMEN AT YALE UNIVERSITY. Jour. Bone and Joint Surg. 4: 247–265, illus.
(19) DAVIS, HORACE B.
   1939. BUSINESS MORTALITY: THE SHOE MANUFACTURING INDUSTRY. Harvard Business Rev. 17: 331–338.
(20) DICKSON, FRANK D., and DIVELEY, REX L.
   1939. FUNCTIONAL DISORDERS OF THE FOOT, THEIR DIAGNOSIS AND TREATMENT. 305 pp., illus. Philadelphia, Montreal, and London.
(21) DUCKWORTH, W. L. H.
   1919. THE INTERNATIONAL AGREEMENT FOR THE UNIFICATION OF ANTHROPOMETRIC MEASUREMENTS TO BE MADE ON THE LIVING SUBJECT. Amer. Jour. Phys. Anthrop. 2: 61–67.
(22) FINLEY, MARY BROUWER.
   1939. HOW TO STIMULATE JUVENILE BUSINESS. Creative Footwear 17 (5): 38, 40, illus.
(23) GILL PUBLICATIONS, INC.
   1930. THREE HUNDRED YEARS OF SHOE AND LEATHER MAKING IN MASSACHUSETTS. 78 pp., illus. Boston.
(24) GOLDING, F. Y., ED.
   1934–35. BOOTS AND SHOES: THEIR MAKING, MANUFACTURING AND SELLING ... 8 v., illus. London.
(25) GRAHAM, W. B.
   1932. BOOTS AND SHOES. Encycl. Amer. 4: 254–261. New York and Chicago.
(26) KNEELAND, HILDEGARD, AND OTHERS.
   1938. CONSUMER INCOMES IN THE UNITED STATES. THEIR DISTRIBUTION IN 1935–36. Natl. Resources Com. 104 pp., illus. Washington, D. C.
(27) MCKAY, A. T.
   1930. LAST MEASUREMENT. Brit. Boot, Shoe and Allied Trades Res. Assoc. Res. Rpt. No. 33, 26 pp., illus. London.

(28) ——— and BRADLEY, H.
   1929. STATISTICAL STUDY OF THE FEET OF THE BOOT AND SHOE OPERATIVES IN NORTHAMPTON. PART I. Brit. Boot, Shoe and Allied Trades Res. Assoc. Res. Rpt. No. 29, 21 pp., illus. London.
(29) ——— and BRADLEY, H.
   1930. STATISTICAL STUDY OF THE FEET OF THE BOOT AND SHOE OPERATIVES IN NORTHAMPTON. PART II. Brit. Boot, Shoe and Allied Trades Res. Assoc. Res. Rpt. No. 31, 33 pp., illus. London.
(30) MONROE, DAY, PENNELL, MARYLAND Y., PHELPS, ELIZABETH, HOPPER, JUNE CONSTANTINE, and HOLLINGSWORTH, HELEN.
   1941. FAMILY EXPENDITURES FOR CLOTHING. FIVE REGIONS. FARM SERIES. U. S. Dept. Agr. Misc. Pub. 428, 387 pp., illus.
(31) MORTON DUDLEY, J.
   1935. THE HUMAN FOOT. ITS EVOLUTION, PHYSIOLOGY AND FUNCTIONAL DISORDERS. 244 pp., illus. New York.
(32) ———
   1939. OH. DOCTOR! MY FEET! 116 pp., illus. New York and London.
(33) MUNSON. EDWARD LYMAN.
   1912. THE SOLDIER'S FOOT AND THE MILITARY SHOE: A HANDBOOK FOR OFFICERS AND NONCOMMISSIONED OFFICERS OF THE LINE. 147 pp., illus. Menasha, Wis.
(34) O'BRIEN, RUTH, PETERSON, ESTHER C., and WORNER, RUBY K.
   1929. BIBLIOGRAPHY ON THE RELATION OF CLOTHING TO HEALTH. U. S. Dept. Agr. Misc. Pub. 62, 146 pp.
(35) ——— and GIRSHICK, MEYER A.
   1939. CHILDREN'S BODY MEASUREMENTS FOR SIZING GARMENTS AND PATTERNS. A PROPOSED STANDARD SYSTEM BASED ON HEIGHT AND GIRTH OF HIPS. U. S. Dept. Agr. Misc. Pub. 365, 25 pp., illus.
(36) ——— GIRSHICK, MEYER A., and HUNT, ELEANOR.
   1941. BODY MEASUREMENTS OF AMERICAN BOYS AND GIRLS FOR GARMENT AND PATTERN CONSTRUCTION. A COMPREHENSIVE REPORT OF MEASURING PROCEDURES AND STATISTICAL ANALYSIS OF DATA ON 147,000 AMERICAN CHLDREN. U. S. Dept. Agr. Misc. Pub. 366, 141 pp., illus.
(37) ——— and SHELTON, WILLIAM C.
   1941. WOMEN'S MEASUREMENTS FOR GARMENT AND PATTERN CONSTRUCTION. U. S. Dept. Agr. Misc. Pub. 454. [In press.]
(38) PENNELL, MARYLAND Y., MONROE, DAY, CRONISTER, KATHRYN, DEPUY, GERALDINE S., and ELLSWORTH, MARJORIE W.
   1941. FAMILY EXPENDITURES FOR CLOTHING. FIVE REGIONS. URBAN AND VILLAGE SERIES. U. S. Dept. Agr. Misc. Pub. 422. [In press.]
(39) PERLMAN, J., JONES, P. L., WITMER, O. R., and ROGERS, H. O.
   1939. EARNINGS AND HOURS IN SHOE AND ALLIED INDUSTRIES DURING FIRST QUARTER OF 1939. BOOTS AND SHOES; CUT STOCK AND FINDINGS; SHOE PATTERNS. U. S. Bur. Labor Statis. Bul. 670, 86 pp.
(40) SCHNITZER, J. G.
   1937. LEATHER FOOTWEAR. WORLD PRODUCTION AND INTERNATIONAL TRADE. U. S. Bur. Foreign and Dom. Com. Trade Prom. Ser. 168, 181 pp., illus.
(41) U. S. BUREAU OF THE CENSUS.
   1940. INDUSTRY CLASSIFICATIONS FOR THE CENSUS OF MANUFACTURES, 1939. 100 pp. Washington, D. C.
(42) U. S. BUREAU OF LABOR STATISTICS.
   1941. FAMILY EXPENDITURES IN SELECTED CITIES, 1935–36. VOL. III, CLOTHING AND PERSONAL CARE. U. S. Dept. Labor Bul. 648, 578 pp., illus.
(43) U. S. DISTRICT COURT, DISTRICT OF MASSACHUSETTS.
   1914. UNITED STATES V. UNITED STATES SHOE MACHINERY COMPANY OF NEW JERSEY ET AL. CLOSING ARGUMENT FOR DEFENDANTS, BY FREDERICK P. FISH, ESQ. 57 pp. Boston.
(44) U. S. FEDERAL TRADE COMMISSION.
   1941. LEATHER BOOT AND SHOE MANUFACTURING CORPORATIONS. Indus. Corp. Rpt., 19 pp.

www.ingramcontent.com/pod-product-compliance
Lightning Source LLC
LaVergne TN
LVHW041311080426
835510LV00009B/964